Assessing the performance of
Phase Change Materials in buildings

Corinne Williams

The research and writing for this publication has been funded by BRE Trust, the largest UK charity dedicated specifically to research and education in the built environment. BRE Trust uses the profits made by its trading companies to fund new research and education programmes that advance knowledge, innovation and communication for public benefit.

BRE Trust is a company limited by guarantee, registered in England and Wales (no. 3282856) and registered as a charity in England (no. 1092193) and in Scotland (no. SC039320). Registered office: Bucknalls Lane, Garston, Watford, Herts WD25 9XX
Tel: +44 (0) 333 321 8811
Email: secretary@bretrust.co.uk
www.bretrust.org.uk

IHS Markit (NYSE: INFO) is the leading source of information, insight and analytics in critical areas that shape today's business landscape. Businesses and governments in more than 165 countries around the globe rely on the comprehensive content, expert independent analysis and flexible delivery methods of IHS Markit to make high-impact decisions and develop strategies with speed and confidence. IHS Markit is the exclusive publisher of BRE publications.

IHS Global Ltd is a private limited company registered in England and Wales (no. 00788737).
Registered office: The Capitol Building, Oldbury, Bracknell, Berkshire RG12 8FZ. www.ihsmarkit.com

BRE publications are available from www.brebookshop.com
or
IHS BRE Press
The Capitol Building
Oldbury
Bracknell
Berkshire RG12 8FZ
Tel: +44 (0) 1344 328038
Fax: +44 (0) 1344 328005
Email: brepress@ihs.com

Printed using FSC or PEFC material from sustainable forests.

FB 84
First published 2016
ISBN 978-1-84806-452-2

Cover images
Right (above): © Phase Energy Ltd
Right (below): © Martyn McLaggan, University of Edinburgh

Contents

Acknowledgements

This publication has been produced as an output of a BRE Trust funded project on Phase Change Materials (PCMs).

The author would like to thank the following people:

- Ian Biggin, Phase Energy Ltd, UK technical representative of BASF (project partner) for helpful assistance and provision of technical information on PCM building products.
- John Hunter, BRE (Fire Sciences and Building Products Passive), about assessment methodologies for reaction to fire performance of building products.
- Richard Jones, BRE (Fire Sciences and Building Products Passive), about testing methodologies for resistance to fire performance of building products.
- Martyn McLaggan, BRE Centre for Fire Safety Engineering, University of Edinburgh, for the contribution about his PhD study on the fire performance of PCMs.
- Kim Allbury and Joanne Munday, BRE (Sustainable Products), about Environmental Product Declarations.
- Debbie Smith, BRE Global, for technical oversight.
- Martyn Webb, BRE (Building Technology), about testing methodologies for building products.

The author is also grateful to the following people for the provision of figures:

- Martyn McLaggan, University of Edinburgh for Figures 9, 10, 11 and 12.
- Ian Biggin, Phase Energy Ltd for Figures 4, 5 and 7.

Executive summary

Phase change materials (PCMs or latent heat storage materials) for building applications are an emerging technology in the UK. PCMs are now becoming available incorporated into different building products; they come in different physical forms and with different transition temperatures. They can provide additional thermal mass to otherwise thermally lightweight buildings currently being constructed and to existing building stock. There is a need to establish sound methodologies for determining the true benefits of these products when installed in a building throughout their life, and for assessing the impact of incorporation of PCMs on broader building performance characteristics.

This publication provides an overview of PCM building products and available methodologies for assessing them. It will be of interest to specifiers, designers, installers, approving authorities, manufacturers, fire safety risk assessors and other interested parties to ensure that this technology is properly considered.

The first part of this publication provides an introduction to the subject of PCM building products. It covers the following subjects: what they are and how they work, their benefits, current technical developments and available products. The second part covers the assessment (test and evaluation methodologies) of PCMs for building applications for long-term thermal performance, environmental impact, structural performance, health and safety considerations, and performance in fire and quality standards.

Glossary

Cycle

Complete melting and re-crystallisation of a substance or material

Environmental profile

A method of identifying and assessing the environmental effects associated with building materials

Hempcrete

A combination of chopped hemp shiv (a renewable biomaterial) and binder comprising of natural hydraulic lime and a small amount of cement

Latent heat storage

Storage and release of thermal energy in materials or substances using the specific latent heat of fusion method

Macro-encapsulated

Involves large volume 'containers'

Micro-encapsulated

Involves dividing the Phase Change Material (PCM) into microscopic, shallow containers, or capsule shells

Phase change

When a material or substance changes from one crystal state to another, eg solid to liquid, liquid to gas

Phase Change Material (PCM)

Latent heat storage material or substance with a high specific latent heat of fusion which, when melting or solidifying at a certain temperature, is capable of storing or releasing large amounts of energy

Phase Change Material (PCM) composite

Phase Change Material (PCM) combined with another material

Sensible heat storage

Storage and release of thermal energy in materials or substances using a specific heat capacity method

Specific heat capacity

Amount of energy needed to raise the temperature of unit mass of that material or substance by 1°C

Specific latent heat of fusion

Amount of energy absorbed when unit mass of that material or substance reaches its melting temperature and changes phase from solid to liquid, at almost constant temperature

Super- (sub- or under-) cooling

When the temperature of the liquid is lowered below its freezing point, without it becoming a solid

Thermal mass

Property of the mass of a building which enables it to store heat, providing 'inertia' against temperature fluctuations

Transition temperature

Temperature at which a material or substance changes from one crystal state to another, eg solid to liquid

1 Introduction

Traditional construction relies on heavyweight structures with high thermal mass to achieve effective energy storage in the construction's fabric. Modern, lightweight structures, with low thermal mass, can be susceptible to overheating[1, 2] due to solar heating and internal heat gains which cannot be absorbed by the structure or being too cold when external temperatures are too low. Air conditioning and heating systems are often specified to make the internal temperatures acceptable.

Phase Change Materials (PCMs), or latent heat storage materials for building applications, are an emerging technology in the UK. They offer a solution for improving the thermal performance and comfort of low thermal mass buildings currently being constructed using modern materials and techniques and also when renovating existing building stock.

PCMs can be used to provide thermal mass to buildings with low thermal mass. They can smooth the daily fluctuations in internal temperatures by lowering the peak temperatures resulting from extreme external temperature changes and can prevent overheating (Figure 1). PCMs can reduce the capacity of mechanical heating or cooling systems.

PCMs come in many physical forms and are available for different transition or operating temperatures. PCMs have now become available and incorporated into different building products.

This publication concentrates on PCMs as part of a passive/fabric/thermal mass approach but it should be noted that PCMs can also be used as part of low energy ventilation and cooling systems and other emerging applications.

The publication provides introductory information about PCM building products and available methodologies for assessing them for specifiers, designers, approving authorities, manufacturers, fire safety risk assessors and other interested parties so they can ensure that this technology is properly considered on individual building projects.

1.1 PCM building products and other temperature control measures

PCM building products, like sensible thermal mass, should be considered as part of the overall package of temperature control measures in a building. These include insulation, shading, building orientation and ventilation. PCMs may not always be the best or only solution and alternative measures should be considered to ensure that the most appropriate form of temperature control measure is used.

It may be advisable for specialist advice to be sought in the early stages when considering the use of PCM and PCM composite building products to ensure their correct application.

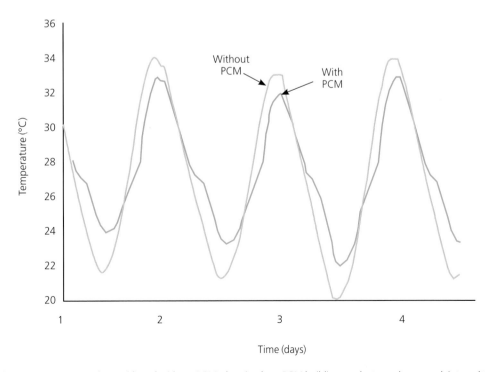

Figure 1: Indoor air temperature versus time, with and without PCM, showing how PCM building products can lower peak internal temperatures

1.2 What are PCMs and how do they work?

PCMs are substances with a high specific latent heat of fusion which, when melting or solidifying at certain temperatures, are capable of storing or releasing large amounts of energy.

Alternatively, PCMs can absorb large amounts of energy by changing phase (eg from solid to liquid) when the ambient temperature reaches a defined level, known as the transition temperature. This energy is released as the ambient temperature drops below the transition temperature and the phase change is reversed.

The most commonly used and well known PCM is ice (Figure 2).

Figure 2: One common application of ice used as a PCM to cool a drink

1.2.1 Sensible and latent heat storage

When heat is applied to a substance, its temperature rises; conversely, when heat is removed from a substance, its temperature decreases. Each substance has a characteristic specific heat capacity, which is the amount of energy needed to raise the temperature of unit mass of that substance by 1°C. The storage and release of thermal energy in materials using the specific heat capacity is termed the sensible heat storage method (Figure 3).

This is expressed in the following equation: $Q = m \cdot c_p \cdot \Delta T$

Where Q is the quantity of heat (in kJ), m is the mass of material (in kg), c_p is the specific heat capacity (in kJ kg^{-1} K^{-1}) and ΔT is the change in temperature (in K).

A PCM can, in addition to the sensible heat method, be used to store and release thermal energy by what is termed the latent heat storage method. This is achieved in practice using the solid to liquid phase change. A PCM absorbs a large amount of latent heat energy, known as the specific latent heat of fusion, when unit mass of the substance reaches its melting temperature and changes phase from solid to liquid, at almost constant temperature; conversely, it releases a large amount of stored latent heat energy when it changes from liquid to solid (Figure 3).

This is expressed in the following equation: $Q = L \cdot m$

Where Q is the energy released or absorbed in the phase change (in kJ), L is the specific latent heat of fusion (in kJ/kg) and m is the mass of the material (in kg).

For PCMs to be effective for thermal energy storage they should possess the following attributes:

- high latent heat of fusion per unit volume
- high thermal conductivity, high specific heat capacity and high density
- sharp melting temperature lying in the practical range of operation
- melt congruently with minimum sub- or under-cooling
- chemically stable
- compatibility with container materials
- non-toxic, non-corrosive, non-flammable/ignition resistant and non-explosive
- low in cost
- abundant.

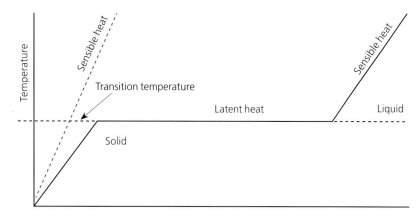

Figure 3: Principles of latent heat storage

1.2.2 Types and properties

Organic and inorganic materials are the most common groups of PCMs. Organic PCMs include paraffin (C_nH_{2n+2}) and fatty acids ($CH_3(CH_2)_{2n}COOH$). Inorganic PCMs include salt hydrates (M_nH_2O).

The advantages and disadvantages of organic and inorganic PCMs are summarised in Table 1.

PCMs come in different physical forms, from wax sheets to inorganic powders, and with different transition temperatures.

Table 1: Comparison of organic and inorganic materials for heat storage[3]

	Organic materials	Inorganic materials
Advantages	– No corrosives – Low or no under-cooling – Chemical and thermal stability	– Greater phase change enthalpy
Disadvantages	– Lower phase change enthalpy – Low thermal conductivity – Flammable	– Super-cooling – Corrosion – Phase separation – Phase segregation, lack of thermal stability

1.2.3 Encapsulation methods

PCMs can be encapsulated in two main ways: macro- and micro-encapsulation.

Macro-encapsulation involves large volume 'containers'. Micro-encapsulation involves dividing the PCM into microscopic, shallow containers, or capsule shells.

The main advantages of encapsulation are: to provide a larger heat transfer area, to reduce reactivity to the outside environment and to control changes in the volume of the storage material as the phase change occurs.

Therefore, the encapsulation material should conduct heat well, be durable enough to withstand frequent changes in the storage material's volume as phase changes occur, restrict the passage of water so that the materials will not dry out and resist leakage and corrosion.

1.3 Benefits of PCMs

Reasons for choosing PCM building products include[4, 5]:

- they provide thermal mass and are particularly suited to lightweight structures where provision of thermal mass is difficult
- there is a choice of transition temperatures
- they have maximum effect over the selected temperature range and behave like a heavyweight structure within that range

- they can replace thermal mass where, for example existing concrete has been decoupled from the room air
- they can be used selectively and are only installed in the most appropriate places within a building.

Barriers to the take-up of PCM building products include[4, 5]:

- flammability of paraffin waxes
- stability/longevity of salt hydrate systems including packaging
- cost effectiveness
- lack of DCLG-approved modelling software[6] including PCMs for calculating the energy performance of buildings
- there are no UK regulations which apply specifically to PCMs themselves (other than chemical and health and safety).

1.3.1 Available paraffin wax-based PCM building products

PCMs are now becoming available in different building products.

As PCM building products are an emerging technology, there is a need to keep abreast of the available PCM building products available for the UK (and European) market and their applications.

Table 2 summarises UK and European paraffin wax-based PCM building products current and recent at the time of writing[a1– i1], with the building application that can be considered (Figure 4). The name and the company are detailed in the references for each product. It has been noted over the past 5 years that a number of new products have been launched and also some have been withdrawn from the UK market, as would be expected for an emerging market.

It can be seen that most of the applications are for ceilings and walls, eg panels, plaster, boards and tiles. These PCM products are for use inside buildings.

Table 2: Paraffin wax-based PCM building products

PCM	Building application
PCM 1[a1]	Lightweight thermal mass panels
PCM 2[b1]	Gypsum-based machine applied plaster (one-layer indoor applications)
PCM 3[c1]	Clay building boards (internal walls and ceilings)
PCM 4[d1]	Dry wall boards
PCM 5[e1]	Ceiling and wall panels
PCM 6[f1]	Plaster
PCM 7[g1]	Clip wall system with prefabricated components
PCM 8[h1]	Acoustic plaster

PCMs 2 to 8 contain micro-encapsulated PCM raw material
PCM 9[i1]

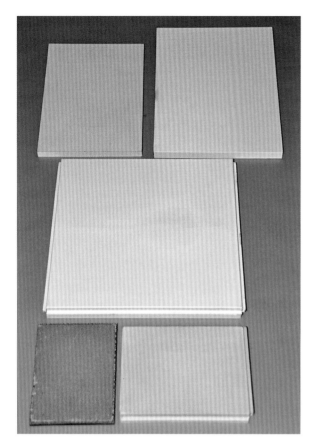

Figure 4: Paraffin wax-based PCM building products.
Image © Phase Energy Ltd

Table 3 lists some of the physical properties of these PCM building products, where possible, which include the melting temperature, weight or area weight, thickness, width and length.

PCM 1

PCM 1 is an aluminium foil-faced sheet material. The macro-encapsulated PCM operates between 18°C and 22°C. The core is a fine mixture of an ethylene-based polymer (40%) and paraffin wax (60%) laminated on both sides with a 100 μm aluminium sheet. A 75 μm thick and 50 mm wide aluminium tape is used to seal the edges and any puncture holes in the laminated face.

Micro-encapsulated PCM raw material and associated products

Micro-encapsulated PCM raw material PCM 9[i1] consists of selected paraffin waxes with an appropriate melting point that have been (micro-) encapsulated in microscopically small, highly cross-linked, acrylic polymer capsules. These capsules can then be incorporated into lightweight building products.

These raw products are listed in the technical literature. PCM 9 is offered as PCM dispersions in water and, for dry mix applications, as re-dispersible powders.

This changes phase within the indoor temperature and human comfort range, ie at 21°C, 23°C and 26°C. The 21°C product is recommended for use in surface cooling systems, the 23°C product for stabilisation of indoor air temperatures (both active and passive applications) and the 26°C product for excessive heating protection (eg lofts and passive protection in warmer regions).

The microcapsules in the aqueous solution are singular within a diameter range of about 2 to 20 μm. The powder form contains larger particles made up of thousands of primary particles bound together to give diameters of 1000 to 3000 μm.

There are a number of construction products that incorporate the PCM micro-encapsulated raw material PCM 9[i1].

PCM 2

PCM 2 is a machine-applied plaster incorporating 20% of the micro-encapsulated PCM raw material PCM 9. The thickness can be varied.

PCM 3

PCM 3 is a drywall, unfired clay, 16 mm thick, building board that can be used for internal walls and ceilings. A 10 mm paper honeycomb structure is laminated onto a 4 mm timber backing, filled with PCM clay material and finished with a 2 mm clay surface. It contains 3 kg/m² of micro-encapsulated PCM raw material PCM 9.

Table 3: Physical properties of current paraffin wax-based PCM building products

PCM	Melting temperature (°C)	Area weight (kg/m²)	Thickness (mm)	Width (mm)	Length (mm)
PCM 1	21.7	4.5	5.26	1000	1198
PCM 2	–	Not applicable	At least a 15 mm coating	Not applicable	Not applicable
PCM 3	23 or 25	15	16	625	1250
PCM 4	23 or 26	25	25	500	1000
PCM 5	23	25	25	600	600
PCM 6	26	–	–	–	–
PCM 7	–	–	–	–	–
PCM 8	23	7.2	12	Not applicable	Not applicable

– denotes not available

PCM 4

PCM 4 is a dry wall board which incorporates 3 kg of PCM/m^2 that can be used in partition walls, interior linings and suspended ceilings.

PCM 5

PCM 5 is a metal ceiling tile with a 24 mm infill incorporating micro-encapsulated PCM raw material PCM 9. The metal cassette is suspended in manufacturer-provided exposed grid systems. The installation can involve a combination of tiles with and without PCM.

PCM 6

PCM 6 is an interior plaster.

PCM 7

PCM 7 is a partition system finished with clay and contains micro-encapsulated PCM raw material PCM 9.

PCM 8

PCM 8 is a plaster made of natural stone, fine mineral aggregates and hydraulic binders which contains PCM 9. It has additional acoustic properties.

Salt hydrate PCM building products

Salt hydrate PCM building products are also available[a2, b2] (Figure 5).

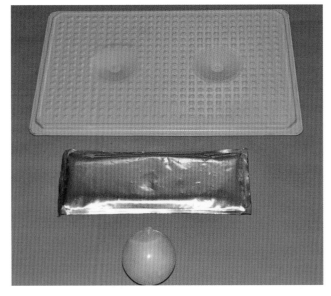

Figure 5: Salt hydrate-based PCM building products. Image © Phase Energy Ltd

1.4 Current developments

1.4.1 Real building performance, experimental studies and modelling tools

The performance of PCM building products installed in real buildings over a long period of time and experimental studies with appropriate instrumentation, and approved predictive modelling tools can provide quantitative data to specifiers, designers and approving authorities to aid their decision making.

Experiments can be carried out to assess the performance of PCM building products in full scale test rooms[7]. The experimental work involves the use of appropriate instrumentation, conducting measurements with and without the PCM, over suitable time periods (Figure 6). This experimental data can be utilised to evaluate theoretical predictions of PCM performance (Figure 7) made with suitable mathematical models.

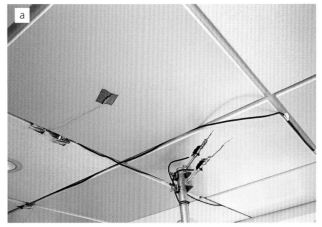

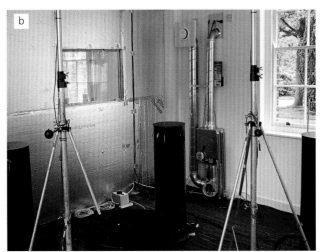

Figure 6: Examples of installed instruments for experimental measurements. (a) Thermocouples measuring temperatures on the surface and embedded in the ceiling tiles containing PCM and (b) Instruments set up inside a room

Figure 7: Specific heat versus temperature for a PCM plasterboard (blue line) and concrete without PCM (red) (calculated data for the PCM/plasterboard is based on measured DSC data for micro-encapsulated PCM by itself; the melting temperature, the peak is 23°C). Image © Phase Energy Ltd

PCM products have been installed in real buildings. The performance of PCM building products can also be monitored in a real building in a similar way to test rooms.

There are a number of demonstration/exemplar buildings in the UK where PCMs have been installed, which include[a3 – d3]:

- the east wing of Somerset House, London, third floor incorporated clay boards containing PCM[a3]
- the visitor centre at the BRE Innovation Park, Ravenscraig, Lanarkshire, which incorporates a PCM ceiling panel system[b3] (Figure 8)
- the Victorian Terrace at BRE Watford, Hertfordshire, which incorporates PCM ceiling tiles[c3]
- the BASF House at the University of Nottingham, which incorporates PCM wall boards[d3].

There are a number of computer tools that are becoming available which incorporate PCMs. Endnotes a4 to f4 give current examples. Some are simple tools but others are complex and require expert users and detailed input data.

PCMs can be modelled in detail using dynamic simulation thermal models incorporating PCM modules. There has been a lack of a DCLG-approved numerical modelling software for PCMs; however, it is understood that one such model is under development[a4].

1.4.2 Other research

PCM building products have and continue to be the subject of various research studies[8, 9, 10, 11]. One example is a BRE Trust funded study of the fire performance of PCM building products discussed in section 1.4.3 (see further reading, McLaggan, 2016).

Figure 8: Visitor centre, BRE Ravenscraig Innovation Park (a) Exterior of building (b) Interior showing ceiling tiles containing PCM

1.4.3 Study on the fire performance of PCM building products

The BRE Trust funded a study of the fire performance of PCM building products[8] to understand how the PCM products behave in realistic fire conditions, what key parameters control this performance and to develop an appropriate robust methodology for assessing their fire performance.

The study has specifically considered how the PCMs modify building materials by casting PCM into the matrix of another suitable product. A hemp-lime composite material was chosen as the substrate so that known amounts of commercially available micro-encapsulated PCM could be incorporated, and their effect evaluated.

The fire performance has been characterised primarily utilising the small-scale cone calorimeter apparatus to study the fire behaviour of PCMs. The specimens were 100 mm by 100 mm and 100 mm thick. The conical shaped heater mounted above the sample exposed the sample to different levels of heat flux. In this study, the sample was sealed at the sides to provide simulated one-dimensional heat transfer. A specific porous sample holder was designed for this study.

A large parametric study has been undertaken which varied the percentage of PCM in the hemp-lime material, the heat flux and repeats. During each test, measurements taken included mass loss rate, oxygen depletion, carbon dioxide and carbon dioxide concentrations, and temperatures in the sample at 10 mm intervals. Downwards smouldering combustion of the specimen was achieved and the specimens were physically examined following each test (Figure 9).

The results indicate that the critical heat flux for smouldering ignition was lowered from 8 kW/m^2 to 3 kW/m^2 when PCMs were added to the hemp-lime material. The peak mass loss increased from 2.8 g/s.m^2 to 5.2 g/s.m^2 when PCMs were added, representing a higher burning rate[9]. It is theorised that an increase in the effective heat of combustion in the PCM case, causes more energy to be released, thereby increasing the temperature and rate of reactions. This, in turn, increases the smoulder spread rate and mass loss rate. The change in behaviour caused by the addition of PCMs has been characterised and quantified.

Thermogravimetric analysis and differential scanning calorimetry has provided information on the thermal degradation steps for PCM and hemp-lime. These results show that the major exothermic oxidation peak for PCMs is lower than hemp-lime, explaining the reduction in critical heat flux in the cone calorimeter testing.

A commercially-available PCM plasterboard product, containing the same micro-encapsulated product used previously, was obtained and also tested in the cone calorimeter to provide a base-line characterisation of an end use product (Figure 10). Samples were 100 mm by 100 mm and 25 mm thick, and tested in both horizontal and vertical orientations.

Results indicate that the critical heat flux for this particular product is 20 kW/m^2 and the time to ignition is 150, 32 and 18 s at 20, 50 and 70 kW/m^2 heat fluxes, respectively, in the horizontal orientation. The peak heat release rate per unit area, based on oxygen consumption, was 59, 74 and 105 kW/m^2 in the horizontal, and 64, 114 and 133 kW/m^2 in the vertical when exposed to 20, 50 and 70 kW/m^2. This knowledge allows fire performance to be used as the goal of the design. As such, it can then be balanced with energy savings to achieve an optimised design which otherwise would not be possible[10].

Figure 9: PCM hemp-lime composite material used for a study on fire performance of PCMs after cone calorimeter testing at 16 kW/m^2. Image © Martyn McLaggan, University of Edinburgh

Figure 10: Cone calorimeter fire testing of commercially available PCM plasterboard tested at 90 kW/m^2. Image © Martyn McLaggan, University of Edinburgh

Lastly, indicative LIFT (lateral flame spread) and Single Burning Item (SBI) tests (BS EN 13823[c8]) with additional detailed temperature measurements inside the samples were conducted at the BRE test facilities (Figures 11 and 12, respectively). The in-depth temperature profiles and results from this work are currently being analysed and will be compared with results from the bench-scale testing. This will enhance our knowledge of the fire behaviour of these types of materials and could be used to develop guidance on how to best optimise their use.

Figure 11: Single burning item (SBI) fire testing (BS EN 13823[c8]) of commercially available PCM with additional sample instrumentation. Image © Martyn McLaggan, University of Edinburgh

Figure 12: IMO spread of flame test apparatus showing testing of commercially available PCM with additional sample instrumentation. Image © Martyn McLaggan, University of Edinburgh

2 Assessment of PCMs and methodologies

There is a need to establish sound methodologies for determining the true benefits of PCM products when installed in a building, throughout their life and for assessing their impact on broader building performance characteristics.

A PCM for building applications will need to perform in accordance with the requirements of the building regulations, associated guidance (Approved Documents for England and Wales, Technical Handbooks for Scotland and Technical Booklets for Northern Ireland) and standards and relevant legislation and directives, including health and safety legislation.

Assessment of PCMs for building applications includes the following topics:

- long-term thermal performance
- environmental impact
- structural performance
- health and safety considerations
- performance in fire
- installation and maintenance.

2.1 Long-term thermal performance

Testing methodologies for the long-term thermal performance of PCMs can be divided into two categories: PCM-specific tests and other building products 'durability' tests.

2.1.1 PCM-specific tests

The current testing specifications[12] used by a German Institute for Quality Assurance and Certification (RAL, www.ral.de) include three tests for PCMs and PCM composites to measure:

- phase change transition temperature and stored heat as function of time
- cycling stability
- thermal conductivity.

Phase change transition temperature and stored heat as a function of time are measured using Differential Scanning Calorimetry (DSC) or other suitable specified calorimetry methods:

- heat flux-DSC dynamic measurement with constant rates of heating and cooling
- heat flux-DSC quasi-stationary measurement with a step profile for heating
- temperature modulated-DSC
- temperature-history method
- CALVET calorimeter.

Thermal conductivity can be measured for the solid and liquid state using the guarded hot plate, hot wire or rod apparatus, as relevant to the PCM product type (Table 4).

Table 4: Measurement method for thermal conductivity of PCMs

Type of sample	Measurement method (solid PCM)	Measurement method (liquid PCM)
Paraffin	Hot wire	Hot wire
Salt hydrates	Guarded hot plate	
Micro-encapsulated PCM		
Composite panel of carbon/PCM	Rod instrument (DIN 51908[13]), hot wire (if isotropic)	
Composite panel of matrix material/micro-encapsulated PCM	Guarded hot plate, hot wire (if isotropic), rod equipment (if thermal conductivity >0.4 W/(mK))	

Table 5: Cycling categories

Cycling category	Number of cycles*	Control measurements for damage criteria every (no. of cycles)
A	≥ 10 000 ~ 30 years	1000 cycles
B	≥ 5000	500 cycles
C	≥ 1000 ~ 3 years	200 cycles
D	≥ 500 ~ 1.5 years	100 cycles
E	≥ 100	20 cycles
F	≥ 50	10 cycles

*One cycle is the complete melting and re-crystallisation.

DIN 51908[13], a German standard, is quoted by RAL for the rod instrument method. BS EN 12667[14] is a British/European standard for the measurement of thermal resistance which uses the guarded hot plate and heat flow meter apparatus.

Thermal cycling to determine stability of the phase transition is carried out in suitable specially-designed cycling equipment which can cycle many times in the shortest possible testing time. One cycle is defined as the complete melting and re-crystallisation of the PCM. The PCM building product needs to survive a defined number of cycles without damage to achieve a particular cycling category (Table 5).

Samples are removed from the cycling apparatus at specified intervals and are measured for the transition temperature and stored heat as a function of temperature and leak tightness (determined visually). The stored heat must not deviate by more than 10% compared to the manufacturer's specification and the phase transition temperature must not deviate more than ± 1 K for onset, peak and offset temperature.

It is a long-term test, eg PCM 9 microcapsules were cycled for >10 000 cycles (cycling category A) at 24 cycles (ie 48 phase changes) per day, which took 16 months to complete[15]. The microcapsules were undamaged and the DSC measurements for the transition temperature and the stored heat were within the required tolerances. The test specifications are described in more detail in reference 12[12].

2.1.2 'Durability' tests for other building products

Selected building products exposure regimes and 'durability' performance tests have been identified as test methodologies relevant to PCM building products.

It is anticipated that PCM building products would be subjected to a particular exposure regime test. A material of known performance, where available, would be exposed at the same time as the PCM building product under test. Expert interpretation of the results of these tests is needed.

The selected exposure regimes are outlined below (also see endnotes as shown).

Freeze-thaw[a5]

A typical freeze-thaw exposure regime consists of freezing a sample in air down to -20°C, holding at this temperature for a number of hours, and then thawing by immersion in water at +20°C.

Heat rain/thermal shock[b5]

A heat rain/thermal shock exposure regime subjects test samples to alternating cycles of radiant heat at typically 60°C to 70°C, followed by cool water spray at between 10°C and 15°C. This method would be most appropriate for products expected to perform in external conditions but may also provide information about a product's ability to withstand thermal shock.

High temperature/temperature cycling[c5]

A temperature exposure/cycling test at above freezing temperatures can be used to replicate situations where a product may be exposed to elevated temperatures for either extended periods at constant exposure or infrequently for a cyclic exposure. For internal PCM products, it may be appropriate to use the temperature range 10°C to 30°C to reflect the typical UK domestic internal temperature range.

Salt spray[d5, e5]

Salt spray exposure can be carried out in different ways, but would generally consist of exposure in a salt laden atmosphere in a salt spray cabinet for metallic components/coatings or cyclic immersion in a sodium chloride solution to mimic the effect of de-icing salts.

Ultraviolet light[f5]

Exposure to ultraviolet light and moisture is a commonly used exposure regime for materials used in an outdoor environment. Most cycles consist of alternating periods of exposure to fluorescent ultraviolet light, followed by exposure to moisture either as a condensation or a water spray. Typical periods of exposure range from 1000 to 5000 h total exposure. Samples can be removed at intervals for examination and testing, if required.

High humidity/moisture

An exposure to high humidity/moisture test can provide valuable performance data on a particular product and how these conditions might affect it. The general approach would be to subject PCM product samples to prolonged periods of high humidity, up to several months. This is a useful test to conduct when assessing how PCM products perform in humid/moist environments, eg bathrooms.

Before and after an exposure regime testing, the PCM sample is measured for general physical properties, for flexural strength, tensile strength, puncture resistance, dimensional stability, as appropriate to see if these have altered. The pre- and post-exposure regime performance tests are as follows (and in endnotes a6 to e6).

Flexural strength[a6]

The calculation of the flexural strength of a sheet specimen is based on measuring the minimum modulus of rupture of the sheet. The sample is mounted on supports and tested once in each direction using a bending test machine. A constant rate of deflection is applied using a loading bar, until the specimen breaks. The modulus of rupture is calculated.

Tensile strength[b6, c6]

To determine the tensile strength of the specimen, the specimen is extended along its major longitudinal axis at constant speed until it fractures or until the stress rod or the strain reaches some pre-determined value. During the procedure, the load sustained by the specimen and the elongation are measured. A universal test apparatus is used for this test.

Puncture resistance[d6]

To determine puncture resistance, the sheet specimen is mounted horizontally on a support and is struck on its top surface by a free falling drop mass with a puncturing tool. Different puncture tools can be selected. The resistance to impact is expressed as the drop height of the puncturing tool which has not caused leakage of the specimen.

Dimensional stability[e6]

To determine dimensional stability, the initial longitudinal and transversal dimensions of the specimen are measured in specified locations. The specimen is then exposed to a regime, eg heating, and the dimensions re-measured.

As the PCM building products identified are for internal use, exposure regimes relevant to external building products PCM applications, ie freeze-thaw, heat rain/thermal shock and ultraviolet, are regarded as low priority.

Corrosion testing, such as salt spray and sulfur dioxide testing, is relevant for PCMs containing metallic components/coatings and if the environment is corrosive, eg factory/industrial environment.

Before and after these exposure regimes it may also be appropriate for the phase change transition temperature and stored heat as a function of time (or cycling stability) to be measured to see if they have changed significantly.

The tests for thermal long-term performance that are most relevant to PCM building products are summarised in Table 6.

Table 6: Long-term thermal performance tests most relevant for PCM building products

Exposure regimes	Performance tests pre- and post- exposure regime	PCM-specific tests pre- and post- exposure regime
High temperature/ temperature cycling[c5]	Flexural strength[a6] Tensile strength[b6, c6] Puncture resistance[d6] Dimensional stability[e6]	Phase change transition temperature and stored heat as function of time[12] (pre- and post-exposure regime)
High humidity/ moisture		Thermal conductivity[12] (one off) Cycling stability[12] (one-off or pre- and post-exposure regime)

2.2 Environmental impact

BRE has prepared draft Product Category Rules (PCR) intended for companies preparing an Environmental Product Declaration for micro-encapsulated organic PCMs used in construction products. The purpose of this publication is to define a consistent approach to the identification, assessment, declaration and communication of the environmental impacts of construction products containing micro-encapsulated organic PCM over its life cycle using life cycle assessment (LCA).

These PCRs include clear guidelines for carrying out the underlying LCA, including collation of data, setting of system boundaries, the environmental impacts to be reported, and the format for communicating the results obtained.

The BRE draft PCR is based on, and represents, a supplement to the European standard EN 15804:2012[16]. The PCR complies with the standard ISO 14044:2006[17], and ISO 14025:2010[18].

It is intended that the draft PCR will be subjected to a consultation process with adequate stakeholder participation and independent third party review before adoption.

2.3 Structural performance

Currently, there are no PCM building products that are structural elements of a building. Therefore, this subject is not discussed further here. If the PCM building product is a structural element of a building then it should comply with the recommendations of Approved Document A: Structure in England[19] and Approved Document A (Structural) in Wales[20], Technical Handbook – Structure[21] in Scotland and Technical Booklet D Structure[22] in Northern Ireland and their relevant supporting performance standards.

2.4 Health and safety considerations

The Workplace (Health, Safety and Welfare) Regulations 1992[23] as amended by the Health and Safety (Miscellaneous amendments and revocations) Regulations 2009[24] contain some requirements which affect building design. The main requirements are covered by the building regulations which apply in England and Wales, Scotland and Northern Ireland.

PCM products for building applications should comply with this and other relevant Health and Safety regulations (eg the Control of Substances Hazardous to Health (CoSHH) Regulations 2004)[25]. The products should have Material Safety Datasheets (MSDS) and CoSHH data sheets. The safety data sheets should comply with UK legal requirements and include all the relevant warnings.

A review of potential health and safety risks of innovative construction products and systems carried out for the Department for Communities and Local Government[5] included PCMs in ceiling applications. The review related to the Building Regulations and therefore, the 'in use phase' of a building.

In terms of toxicity and health hazards, this study found that current PCM building products in ceiling applications are not known to pose any risks to health. The current UK and European micro-encapsulated PCMs use acrylic encapsulation shells. However, if melamine formaldehyde were to be used instead (which could be the case for imported micro-encapsulated PCMs), this would pose a high risk to health as formaldehyde is a known carcinogen[26].

Where the use of PCMs is under consideration, the designer/ specifier should consider any toxic and health hazards by carrying out a review of MSDS and a manufacturer's other technical information for PCMs. As part of the study reported in this publication a number of material safety sheets and manufacturer's technical information where possible, have been reviewed and are briefly summarised below for illustrative purposes.

One company stated that there were other highly efficient materials that could be used as PCMs but that these had been discarded based on safety considerations.

Based on a review of manufacturer's information for the paraffin wax-based PCMs:

- The MSDS for PCM 1 states that no particular hazard has been identified from the PCM panel in normal conditions (except for possible cuts).
- The MSDS for PCM 9 microcapsules states that this product is not irritating to eyes and skin and that a chemical safety assessment is not required.

Most inorganic salt PCMs are stated to be irritant, but non-carcinogenic and non-toxic.

2.5 Performance in fire

2.5.1 Assessing fire performance of PCMs

A PCM product for building applications should also meet the technical recommendations of Approved Document B (Fire safety) in England[27] and Approved Document B (Fire safety) in Wales[28], Technical Handbook – Fire in Scotland[29] and Technical Booklet E – Fire safety[30] in Northern Ireland, and their relevant supporting performance standards. For example, performance standards are covered in Appendix A, of Approved Document B (Fire safety) for England[27] and Approved Document B (Fire safety) for Wales[28].

2.5.2 Assessment of fire resistance of PCMs

Fire resistance is the ability of an element of structure of a building to satisfy for a stated period of time, some or all of the appropriate criteria specified in the relevant test standard.

Elements of a building construction such as wall, floor, door and ceiling systems are required to have a fire resistance rating so that they provide compartmentation to satisfy building regulations. PCMs may be part of such systems, eg contained in a board forming the face of a partition. Fire resistance ratings are only assigned to the system, the end use element of construction, eg a wall system, not to a component, eg a single PCM board.

Performance in terms of fire resistance to be met by construction products is determined by European Commission Decision 2000/367/EC implementing the Council Directive 89/106/EEC[31] with regards to the classification of the resistance to fire performance of the construction products.

Fire resistance performance is determined by national or European tests. Test methods potentially relevant to PCM building products include:

Method for determination of the fire resistance of elements of construction (general principles). BS 476-20:1987[a7]

Method for determination of the fire resistance of loadbearing elements of construction. BS 476-21:1987[b7]

Method for determination of the fire resistance of non-loadbearing elements of construction. BS 476-22:1987[c7]

Method for determination of the contribution of components to the fire resistance of a structure. BS 476-23:1987[d7]

Method for determination of the fire resistance of ventilation ducts. BS 476-24:1987[e7]

Fire resistance tests for non-loadbearing walls. BS EN 1364-1:2015[f7]

Fire resistance tests for non-loadbearing ceilings. BS EN 1364-2:1999[g7]

Fire resistance tests for loadbearing walls. BS EN 1365-1:2012[h7]

Fire resistance tests for loadbearing ceilings. BS EN 1365-2:2000[i7]

Test methods for determining the contribution to the fire resistance of structural members – Horizontal protective membranes. BS EN 13381-1:2014[j7].

Table A1 in Approved Document B (Fire safety)[27, 28], Appendix A, Performance of materials, products and structures; Annex 2.A in Table 2.7 in Technical Handbook Domestic – Fire[29]; Annex 2.D in Table 2.19 in Technical Handbook Non domestic – Fire[29] and Table 4.1 in Technical Booklet E – Fire safety[30] give the specific requirements for each element in terms of performance criteria.

In the European system, the results of fire resistance tests on systems are classified as 'R' (load bearing capacity or resistance to collapse), 'E' (integrity or resistance to fire penetration) and 'I' (insulation or resistance to the transfer of excessive heat), in accordance with BS EN 13501-2[32].

In addition, EOTA ETAG *Guideline for European Technical Approval 018 Part 4 Fire protective products. Fire protective board, slab and mat products and kits*[33] give the essential requirements for board products.

Fire protective boards may be used in many end-use conditions and it is economically impossible to test the boards under consideration for every conceivable end-use condition. Indicative fire resistance testing is advisable in the first instance.

2.5.3 Assessment of the resistance of roofs to external fire exposure

Performance in terms of the resistance of roofs to external fire exposure is determined by either national or European tests and corresponding classification rules. However, currently there are no PCM building products that are used in roof systems, or as coverings, and this is unlikely to change. Therefore, this subject is not discussed further here.

2.5.4 Assessment of reaction to fire performance of PCMs

Performance in terms of reaction to fire to be met by construction products is determined by European Commission Decision 2000/147/EC implementing the Council Directive 89/106/EEC[34] with regard to the classification of the reaction to fire performance of the construction product.

The relevant European test methods for reaction to fire of building products are specified as follows:

Non-combustibility test. BS EN ISO 1182:2010[a8]

Determination of the heat of combustion. BS EN ISO 1716:2010[b8]

Building products excluding flooring exposed to the thermal attack by a single burning item. BS EN ISO 13823:2010 + A1:2014[c8]

Ignitability of building products when subjected to direct impingement of a flame –Single-flame source test. BS EN ISO 11925-2:2010[d8]

Conditioning procedures and general rules for selection of substrates. BS EN ISO 13238:2010[e8].

2.5.5 Non-combustible materials

Non-combustible materials are defined in Table A6 of Approved Document B (Fire safety)[28] and Annex 2.B Table 2.8 in Technical Handbook Domestic – Fire[29].

Annex 2.E Table 2.20 in Technical Handbook Domestic – Fire[29] and in Section 1 of Technical Booklet E – Fire safety[30] either as listed products, or in terms of performance which is determined in accordance with the national or European fire test methods[a9, b9, c9, a8, b8]:

- national class when tested to BS 476-4:1970[a9] and BS 476-11:1982[b9] which is a fire test on building materials and structures. The latter test method includes assessing the heat emission from building materials
- European class when classified as a class A1 in accordance with BS EN 13501-1:2007 + A1:2009[c9]. In order to achieve a classification, data from BS EN ISO 1182:2010[a8] and BS EN ISO 1716:2010[b8] have to be used.

2.5.6 Materials of limited combustibility

Materials of limited combustibility are defined by performance as determined using the national or European fire test methods:

- national class by reference to the method specified in BS 476-11:1982[b9]
- European class in terms of performance when classified as class A2-s3, d2 in accordance with BS 13501-1:2007 + A1:2009[c9]. A classification can be obtained using data from BS EN ISO 1182:2010[a8] or from data from BS EN ISO 1716:2010[b8] and BS EN ISO 13823:2010 + A1:2014[c8].

2.5.7 Internal linings

The contribution of a wall or ceiling surface to the early stages of development of a fire is controlled by defining performance classes which are dependent on the intended end use application and location of a lining system in a building. Under European classifications, lining systems are classified in accordance with BS EN 13501-1:2007 + A1:2009[c9], where class A1 is the best and class F is the worst reaction to fire classification.

Under national classification, surface spread of flame performance is determined in accordance with reference to the BS 476-7:1997[d9] under which materials or products are classified.

Maximum acceptable fire propagation indices are specified to restrict the use of materials which ignite easily, which have a high rate of heat release and/or reduce the time to flashover. These are determined by reference to the method specified in BS 476-6:1989 + A1:2009[e9].

2.5.8 Fire test methods

BS 476-10:2008[35] provides a guide to the various test methods in BS 476.

Existing fire test methods and apparatuses appear to be suitable but the performance criteria may need adapting to cover all types of PCM building products.

2.5.9 PCM building products fire performance test results

Organic PCMs such as paraffin wax are flammable. However, this fire risk can be moderated by protecting the PCM from direct flame attack, eg by macro-encapsulating the paraffin wax-based PCM between two aluminium sheets and sealing the edges with aluminium tape and then installing it behind a board of limited combustibility, such as plasterboard; or by micro-encapsulating paraffin wax-based PCM inside inert spherical shells, and dispersing them throughout a material of limited combustibility.

Examples of indicative test results relating to the reaction to fire performance of a range of PCM products (where the manufacturer's technical data sheets are available), show a broad range of performances from class B – s1,d0 to class E.

2.6 Installation and maintenance of PCM building products

It is important that the manufacturer's product installation and maintenance guidelines are followed to ensure the correct performance of the PCM building product.

2.6.1 Certification schemes

Third party product certification provides a means of identifying products which have demonstrated that they provide a requisite level of performance against appropriate tests and standards. They also provide confidence that the products supplied to the market are to the same specification as that tested and assessed.

In the UK, BRE Global provides independent, third-party certification of fire, security and environmental products and services. The Loss Prevention Certification Board (LPCB) operates certification schemes related to the fire performance of products. These are listed in the LPCB *List of Approved Fire and Security Products and Services*, known as the Red Book[36]. One of the certification schemes relates to environmental profiles – a method of identifying and assessing the environmental effect associated with building materials. These are listed in GreenBookLive[37], a free online database designed to help specifiers and end users identify products and services that can help to reduce their environmental impacts. The database also includes products that are assessed under our rigorous Environmental Profiles and Responsible Sourcing schemes.

3 Quality schemes for PCM-specific attributes

An example of a typical quality scheme that exists for PCM-specific attributes is run by RAL. The RAL quality and testing regulations for PCMs[38] and the quality and testing specifications for PCM and PCM composites[12] can be found on the RAL website (www.ral.de). These are particularly concerned with controlling the definitive properties of PCMs and PCM composites in terms of their energy performance:

- phase transition temperature and stored heat (on heating and cooling)
- cycling stability of the phase transition (defined number of cycles without damage), see cycling categories in Table 5
- thermal conductivity (of solid and liquid)
- product data sheets
- safety data sheets.

The PCM-specific tests have been described earlier, see section 2.1.1 PCM-specific tests.

As part of this particular scheme, product data sheets must be prepared for each product containing PCMs and must include information on the type of product, the operating range, the maximum permissible temperature, the specific weight, the phase transition temperature and stored heat, the reproducibility of the phase transition and the thermal conductivity.

Additionally, the MSDS must comply with the legal requirements of the intended country of sale and include all the relevant warnings.

4 Conclusions and recommendations

4.1 Conclusions

PCM building products used for latent heat storage is an emerging market in the UK and they are being considered to provide thermal mass to lightweight structures and for refurbishment projects; specialist knowledge of the properties of the PCM building product, and its intended application, is required for successful design and implementation.

Where used, they should be considered as part of the overall package of temperature control measures for a building and not only in isolation. They may not always be the best or only solution; alternative measures should be considered for each individual building to ensure that the most appropriate form of temperature control measure is selected.

Various PCM building products have become available for use in buildings; each of these needs to perform in accordance with the requirements of the building regulations, associated guidance (Approved Documents) and standards and comply with relevant legislation and directives, including health and safety legislation. The available information indicates that current paraffin wax melamine formaldehyde-free PCM building products are not known to pose any risks to health in normal use; most inorganic salt PCMs are irritant, but non-carcinogenic and non-toxic.

There are suitable test methodologies for PCM building products for long-term thermal and fire performance; they should be tested against all the relevant tests and achieve the correct classifications. Current products appear to be for internal, non-structural applications.

Manufacturer's installation and maintenance guidelines for PCM building products exist.

PCM building products installed in real buildings over a long period of time and experimental studies, with appropriate measurements, can provide quantitative performance data. Simple and complex computer tools with PCM modules are becoming available (there has been a lack of an 'approved' dynamic simulation model) for PCMs; however, it is understood that the development and approval of this type of model may be underway.

Quality standards that cover the special PCM attributes and third party certification schemes that cover the environmental impact of building products are in place.

4.2 Recommendations

- Specifiers and designers may need to seek specialist advice in the early stages when considering and selecting PCM building products to ensure their correct application and optimal use for a particular building.
- Specifiers and designers should specify that the selected PCM building product is free from melamine formaldehyde to avoid a potentially high health risk.
- Specifiers, designers and approving authorities should request relevant technical evidence from PCM suppliers to aid their decision making to assess:
    - the performance of particular PCM building products in achieving the requirements of each of the relevant test specifications and compliance with regulations, legislation and directives
    - how particular PCM building products, as installed in the real building, will perform as part of the whole package of temperature control measures. This evidence includes: product and MSDS, third party test reports, experimental data and modelling predictions from approved, validated models.
- Specifiers, designers and approving authorities should consider the use of third party approved/certificated PCM building products where possible covering PCM-specific attributes and environmental impact.
- Manufacturers should ensure that they have all the relevant technical information for their PCM and provide this to specifiers, designers and approving authorities for their consideration.
- Installers of PCM building products should follow manufacturer's guidelines for installation and maintenance.
- Fire risk assessors need to be aware that it is very difficult (almost impossible) to visually assess whether a building product contains PCM unless it is marked or labelled; the only way to know this is by being informed and via documentation.

5 References and further reading

References

1. Nicol F and Spires B. The limits of thermal comfort: avoiding overheating in European buildings. TM52. London, CIBSE, 2013.

2. Dengel A and Swainson M. Overheating in new homes. A review of the evidence. NF 46. Milton Keynes, NHBC Foundation, 2012.

3. Zalba B, Marín J M, Cabeza L F and Mehling H. Review of thermal energy storage with phase change: materials, heat transfer, analysis and applications. Applied Thermal Engineering, 2003, 23: 251-283.

4. Biggin I. Soaking up the heat – Phase Change Materials in construction. RIBA assessed CPD, November 2014. Private communication.

5. Department for Communities and Local Government (DCLG). Review of potential health and safety risks of innovative construction products and systems. London, DCLG.

6. DCLG. DCLG approved national calculation methodologies and software programs for buildings other than dwellings (for the production of Energy Performance Certificates, Display Energy Certificates and Air Conditioning Inspection Reports). London, DCLG. December 2012, last updated 31 October 2014.

7. Kuznik F and Virgone J. Experimental assessment of a Phase Change Material for wall building use. Applied Energy, 2009, 86: 2038-2046.

8. McLaggan M S, Hadden R M and Gillie M. Smouldering characteristics of Phase Change Materials when exposed to radiant heat flux. 35th International Symposium on Combustion, San Francisco, USA, 3–8 August, 2014.

9. McLaggan M S, Hadden R M and Gillie M. Smouldering combustion of phase change materials within a porous hemp matrix. 9th Mediterranean Combustion Symposium, Rhodes, Greece, 7–11 June, 2015.

10. McLaggan M S, Hadden R M and Gillie M. Fire performance of plasterboard containing Phase Change Materials. 2nd European Symposium on Fire Safety Science, Nicosia, Cyprus, 16–18 June, 2015.

11. Kontogeorgos D A, Mandilaras I D and Founti M A. Fire behavior of regular and latent heat storage gypsum boards. Fire and Materials, published online in Wiley Online Library, DOI 10.1002/fam.2246. First published online in March 2014. Berlin, Beuth Verlag, 2014.

12. RAL. Phase Change Material. RAL-GZ 896. Quality and test rules. Bonn, RAL, 2013. Available from www.pcm-ral.de.

13. Deutsches Institut fur Normung E.V. (DIN). DIN 51908: Testing of carbon materials – Determination of thermal conductivity at room temperature by means of a comparative method – Solid materials. Berlin, Beuth Verlag, 2006.

14. BSI. Thermal performance of building materials and products - Determination of thermal resistance by means of guarded hot plate and heat flow meter methods – Products of high and medium thermal resistance. BS EN 12667:2001. London, BSI, 2001.

15. Information on Micronal cycling. Available from www.micronal.de/portal/streamer?fid=309980.

16. BSI. Sustainability of construction works. Environmental product declarations. Core rules for the product category of construction products. BS EN 15804:2012. London, BSI, 2012.

17. ISO. Environmental management – Life cycle assessment – Requirements and guidelines). ISO 14044:2006. London, BSI, 2006.

18. ISO. Environmental labels and declarations – Type III environmental declarations – Principles and procedures. ISO 14025:2010. London, BSI, 2010.

19. DCLG. The Building Regulations (England and Wales) 2010. Approved Document A: Structure, 2004 edition, incorporating 2004, 2010 and 2013 amendments. London, DCLG, 2013.

20. Welsh Government. The Building Regulations 2010. Approved Document A (Structural) volume 1, 2004 edition incorporating 2010 amendment. Cardiff, Welsh Government, 2016.

21. Scottish Government. Building Regulations (Scotland) 2004. Technical Handbooks Domestic and Non domestic – Structure. Edinburgh, Scottish Government, 2015.

22. Department of Finance and Personnel. Building Regulations (Northern Ireland) 2012 Technical Booklet D Structure. Belfast, Department of Finance and Personnel, 2012.

23. The Workplace (Health, Safety and Welfare) Regulations 1992.

24. The Health and Safety (Miscellaneous amendments and revocations) Regulations 2009.

25. Control of Substances Hazardous to Health (CoSHH) Regulations 2004.

26. World Health Organization. International Agency for Research on Cancer (IARC) list. IARC monographs on the evaluation of carcinogenic risks to humans. Volume 88. Lyon, IARC, 2006.

27. DCLG. The Building Regulations 2010 (England). Approved Document B (Fire safety) volume 1 Dwellinghouses, and volume 2 Buildings other than dwellinghouses. 2006 editions incorporating 2010 and 2013 amendments. London, DCLG, 2013.

28. Welsh Government. The Building Regulations 2010. Approved Document B (Fire safety) volume 1 Dwellinghouses (for use in Wales), 2007 edition incorporating 2010 and 2016 amendments. Cardiff, Welsh Government, 2016.

29. Scottish Government. Building Regulations (Scotland) 2004. Technical Handbooks Domestic and Non domestic – Fire. Edinburgh, Scottish Government, 2015.

30. Department of Finance and Personnel. Building Regulations (Northern Ireland) 2012. Technical Booklet E – Fire safety. Belfast, Department of Finance and Personnel, 2012.

31. European Commission Decision 2000/367/EC of 3rd May 2000 implementing the Council Directive 89/106/EEC. Available from http://eur-lex.europa.eu.

32. BSI. Fire classification of construction products and building elements. Classification using data from fire resistance tests, excluding ventilation services. BS EN 13501-2:2016. London, BSI, 2016.

33. EOTA. ETAG Guideline for European Technical Approval 018 Part 4 Fire protective products. Fire protective board, slab and mat products and kits. Brussels, EOTA, 2011.

34. European Commission Decision 2000/147/EC of 8th February 2000 implementing the Council Directive 89/106/EEC. Available from http://eur-lex.europa.eu.

35. BSI. Fire tests on building materials and structures. Guide to the principles, selection, role and application of fire testing and their outputs. BS 476-10. London, BSI, 2008.

36. BRE Global. Loss Prevention Certification Board (LPCB). List of approved fire and security products and services (Red Book). Watford, BRE Global. Available at www.redbooklive.com.

37. BRE Global. Reference source and online listing of environmental products and services (Green Book). Watford, BRE Global. Available at GreenBookLive.com.

38. RAL. Detailed Regulations of the RAL-GZ 896 Quality and test rules for PCMs. Available from www.pcm-ral.de.

Further reading

McLaggan, M S. Novel fire testing frameworks for Phase Change Materials and hemp-lime insulation. Doctor of Philosophy thesis. Edinburgh, The University of Edinburgh, 2016. Available at www.era.lib.ed.ac.uk/handle/1842/15896.

6 Endnotes

1 Paraffin wax-based PCMs

a1   Energain, DuPont.

b1   Weber.mur clima, Weber St-Gobain.

c1   Ebb™PCM, Eco Building Boards.

d1   Alba®balance, Rigips Saint-Gobain.

e1   CoolZone, Armstrong World Industries Ltd.

f1   Klima-544 Tynk, Termo Organika.

g1   k.Wand, Alois Scheicher GmbH.

h1   Clima-Akustikputz, Scherff GmbH&CoKG.

i1   Micronal®, BASF.

2 Salt hydrates PCM examples

a2   Climsel™, Climator.

b2   Thermusol K block, Salca BV.

3 Demonstration/exemplar buildings including PCMs

a3   BASF. Reducing CO_2 in buildings, Energy Efficient Somerset House, London.

b3   BASF. Reducing CO_2 in buildings, Visitors centre, BRE Ravenscraig Innovation Park.

c3   BASF. Reducing CO_2 in buildings, The Victorian Terrace, BRE Watford.

d3   BASF. Reducing CO_2 in buildings. The BASF House, University of Nottingham.

4 Planning and simulation programmes for the use of PCMs*

a4   TAS dynamic thermal simulation model, Environmental Design Solutions Limited (EDSL).

b4   PCMexpress Calculation tool for buildings with PCMs, user manual.

c4   TRNSYS Calculation tool for the transient simulation of thermal systems, TRANSSOLAR Energietechnik GmbH.

d4   ESP-r, University of Strathclyde.

e4   WUFI 5, PC-Program for calculating the coupled heat and moisture transfer in building components, Fraunhofer-Institut für Bauphysik IBP.

e4   BASF Micronal® PCM App available from www.micronal.de.

5 Exposure regimes

a5   Natural stone test methods – Determination of frost resistance. BS EN 12371:2010. London, BSI, 2010.

b5   Products and systems for the protection and repair of concrete structures – Test methods – Determination of thermal compatibility – Part 2: Thunder-shower cycling (thermal shock). BS EN 13687-2:2002. London, BSI, 2002.

c5   Flexible sheets for waterproofing – Bitumen, plastic and rubber sheets for roofing – Method of artificial ageing by long term exposure to elevated temperature. BS EN 1296:2001. London, BSI, 2001.

d5   Corrosion tests in artificial atmospheres – Salt spray tests. BS EN ISO 9227:2012. London, BSI, 2012.

e5   Products and systems for the protection and repair of concrete structures – Test methods – Determination of thermal compatibility – Part 1: Freeze-thaw cycling with de-icing salt immersion. BS EN 13687-1:2002. London, BSI, 2002.

f5   Plastics – Methods of exposure to laboratory light sources – Part 3: Fluorescent UV lamps. BS EN ISO 4892-3:2016. London, BSI, 2016.

6 Pre- and post-exposure regimes performance tests

a6   Fibre cement flat sheets – Product specification and test methods. BS EN 12467:2012 + A1:2016. London, BSI, 2012.

b6   Plastics – Determination of tensile properties – Part 1: General principles. BS EN ISO 527-1:2012. London, BSI, 2012.

c6   Plastics – Determination of tensile properties – Part 3: Test conditions for films and sheets. BS EN ISO 527-3:1996. London, BSI, 1996.

d6   Flexible sheets for waterproofing – Bitumen, plastic and rubber sheets for roof waterproofing – Determination of resistance to impact. BS EN 12691:2006. London, BSI, 2006.

e6   Flexible sheets for waterproofing – Determination of dimensional stability – Part 2: Plastic and rubber sheets for roof waterproofing. BS EN 1107-2:2001. London, BSI, 2001.

* See www.pcm-ral.de.

7 Fire resistance test methods: Fire tests on building materials and structures

a7   Fire tests on building materials and structures – Method for determination of the fire resistance of elements of construction (general principles). BS 476-20:1987. London, BSI, 1987.

b7   Methods for determination of the fire resistance of loadbearing elements of construction. BS 476-21:1987. London, BSI, 1987.

c7   Method for determination of the fire resistance of non-loadbearing elements of construction. BS 476-22:1987. London, BSI, 1987.

d7   Methods for determination of the contribution of components to the fire resistance of a structure. BS 476-23:1987. London, BSI, 1987.

e7   Method for determination of the fire resistance of ventilation ducts. BS 476-24:1987. London, BSI, 1987.

f7   Fire resistance tests for non-loadbearing elements – Walls. BS EN 1364-1:2015. London, BSI, 2015.

g7   Fire resistance tests for non-loadbearing elements – Ceilings. BS EN 1364-2:1999. London, BSI, 1999.

h7   Fire resistance tests for loadbearing elements – Walls. BS EN 1365-1:2012. London, BSI, 2012.

i7   Fire resistance tests for loadbearing elements – Floors and roofs. BS EN 1365-2:2014. London, BSI, 2014.

j7   Test methods for determining the contribution to the fire resistance of structural members – Horizontal protective membranes. BS EN 13381-1:2014. London, BSI, 2014.

8 European reaction to fire test methods

a8   Reaction to fire tests for building products – Non-combustibility test. BS EN ISO 1182:2010. London, BSI, 2010.

b8   Reaction to fire tests for building products – Determination of the heat of combustion. BS EN ISO 1716:2010. London, BSI, 2010.

c8   Reaction to fire tests for building products – Building products excluding flooring exposed to the thermal attack by a single burning item. BS EN ISO 13823:2010 + A1:2014. London, BSI, 2010.

d8   Reaction to fire tests – Ignitability of building products when subjected to direct impingement of a flame. Single-flame source test. BS EN ISO 11925-2:2010. London, BSI, 2010.

e8   Reaction to fire tests for building products – Conditioning procedures and general rules for selection of substrates. BS EN ISO 13238:2010. London, BSI, 2010.

9 National reaction to fire tests

a9   Fire tests on building materials and structures – Non-combustibility test for materials. BS 476-4:1970. London, BSI, 1970.

b9   Fire tests on building materials and structures – Method for assessing the heat emission from building materials. BS 476-11:1982. London, BSI, 1982.

c9   Fire classification of construction products and building elements – Classification using test data from reaction to fire tests. BS EN 13501-1:2007 + A1:2009. London, BSI, 2009.

d9   Fire tests on building materials and structures – Method of test to determine the classification of the surface spread of flame of products under which materials of products. BS 476-7:1997. London, BSI, 1997.

e9   Fire tests on building materials and structures – Method of test for fire propagation for products. BS 476-6:1989 + A1:2009. London, BSI, 1989.

Publications
from IHS BRE Press

Fire safety and security in retail premises. **BR 508**

Automatic fire detection and alarm systems. **BR 510**

Handbook for the structural assessment of large panel system (LPS) dwelling blocks for accidental loading. **BR 511**

Design fires for use in fire safety engineering. **FB 29**

Ventilation for healthy buildings: reducing the impact of urban pollution. **FB 30**

Financing UK carbon reduction projects. **FB 31**

The cost of poor housing in Wales. **FB 32**

Dynamic comfort criteria for structures: a review of UK standards, codes and advisory documents. **FB 33**

Water mist fire protection in offices: experimental testing and development of a test protocol. **FB 34**

Airtightness in commercial and public buildings. 3rd edn. FB 35

Biomass energy. **FB 36**

Environmental impact of insulation. **FB 37**

Environmental impact of vertical cladding. **FB 38**

Environmental impact of floor finishes: incorporating The Green Guide ratings for floor finishes. **FB 39**

LED lighting. **FB 40**

Radon in the workplace. 2nd edn. **FB 41**

U-value conventions in practice. **FB 42**

Lessons learned from community-based microgeneration projects: the impact of renewable energy capital grant schemes. **FB 43**

Energy management in the built environment: a review of best practice. **FB 44**

The cost of poor housing in Northern Ireland. **FB 45**

Ninety years of housing, 1921–2011. **FB 46**

BREEAM and the Code for Sustainable Homes on the London 2012 Olympic Park. **FB 47**

Saving money, resources and carbon through SMARTWaste. **FB 48**

Concrete usage in the London 2012 Olympic Park and the Olympic and Paralympic Village and its embodied carbon content. **FB 49**

A guide to the use of urban timber. **FB 50**

Low flow water fittings: will people accept them? **FB 51**

Evacuating vulnerable and dependent people from buildings in an emergency. **FB 52**

Refurbishing stairs in dwellings to reduce the risk of falls and injuries. **FB 53**

Dealing with difficult demolition wastes. **FB 54**

Security glazing: is it all that it's cracked up to be? **FB 55**

The essential guide to retail lighting. **FB 56**

Environmental impact of metals. **FB 57**

Environmental impact of brick, stone and concrete. **FB 58**

Design of low-temperature domestic heating systems. **FB 59**

Performance of photovoltaic systems on non-domestic buildings. **FB 60**

Reducing thermal bridging at junctions when designing and installing solid wall insulation. **FB 61**

Housing in the UK. **FB 62**

Delivering sustainable buildings. **FB 63**

Quantifying the health benefits of the Decent Homes programme. **FB 64**

The cost of poor housing in London. **FB 65**

Environmental impact of windows. **FB 66**

Environmental impact of biomaterials and biomass. **FB 67**

DC isolators for photovoltaic systems. **FB 68**

Computational fluid dynamics in building design. **FB 69**

Design of durable concrete structures. **FB 70**

The age and construction of English homes. **FB 71**

A technical guide to district heating. **FB 72**

Changing energy behaviour in the workplace. **FB 73**

Lighting and health. **FB 74**

Building on fill: geotechnical aspects. 3rd edn. **FB 75**

Changing patterns in domestic energy use. **FB 76**

Embedded security. **FB 77**

Performance of exemplar buildings in use. **FB 78**

Designing out unintended consequences when applying solid wall insulation. **FB 79**

Applying Fabric First principles. **FB 80**

The full cost of poor housing. **FB 81**

The cost-benefit to the NHS arising from preventative housing interventions. **FB 82**

Measuring fuel poverty. **FB 83**

Assessing the performance of Phase Change Materials in buildings. **FB 84**

Material resource efficiency in construction: supporting a circular economy. **FB 85**

Managing risks in existing buildings: an overview of UK risk-based legislation for commercial and industrial premises. **FB 86**

For a complete list of IHS BRE Press publications visit www.brebookshop.com